Science
Grades 3 and up

~A BINGO BOOK~

Animal Characteristics Bingo Book

COMPLETE BINGO GAME IN A BOOK

I0000345

Written By Rebecca Stark

The purchase of this book entitles the individual teacher to reproduce copies of the student pages for use in his or her classroom exclusively. The reproduction of any part of the work for an entire school or school system or for commercial use is prohibited.

ISBN 978-0-87386-443-5

© 2016 Barbara M. Peller

Educational Books 'n' Bingo

Printed in the U.S.A.

ANIMAL CHARACTERISTICS BINGO DIRECTIONS

INCLUDED:

List of Terms

Templates for Additional Terms and Clues

2 Clues per Term

30 Unique Bingo Cards

Markers

1. **Either cut apart the book or make copies of ALL the sheets. You might want to make an extra copy of the clue sheets to use for introduction and review. Keep the sheets in an envelope for easy reuse.**

2. Cut apart the call cards with terms and clues.

3. Pass out one bingo card per student. There are enough for a class of 30.

4. Pass out markers. You may cut apart the markers included in this book or use any other small items of your choice.

5. Decide whether or not you will require the entire card to be filled. Requiring the entire card to be filled provides a better review. However, if you have a short time to fill, you may prefer to have them do the just the border or some other format. Tell the class before you begin what is required.

6. There are 50 terms. Read the list before you begin. If there are any terms that have not been covered in class, you may want to read to the students the term and clues before you begin.

7. There is a blank space in the middle of each card. You can instruct the students to use it as a free space or you can write in answers to cover terms not included. Of course, in this case you would create your own clues. (Templates provided.)

8. Shuffle the cards and place them in a pile. Two or three clues are provided for each term. If you plan to play the game with the same group more than once, you might want to choose a different clue for each game. If not, you may choose to use more than one clue.

9. Be sure to keep the cards you have used for the present game in a separate pile. When a student calls, "Bingo," he or she will have to verify that the correct answers are on his or her card AND that the markers were placed in response to the proper questions. Pull out the cards that are on the student's card keeping them in the order they were used in the game. Read each clue as it was given and ask the student to identify the correct answer from his or her card.

10. If the student has the correct answers on the card AND has shown that they were marked in response to the *correct questions,* then that student is the winner and the game is over. If the student does not have the correct answers on the card OR he or she marked the answers in response to *the wrong questions,* then the game continues until there is a proper winner.

11. If you want to play again, reshuffle the cards and begin again.

Have fun!

© **Barbara M. Peller**

TERMS

adaptation	herbivores
amphibians	insect
Animalia	invertebrate(s)
annelids (Annelida)	kingdom
arachnids (Arachnida)	larva (larvae)
arthropods (Arthropoda)	mammals (Mammalia)
biped(s)	marsupials
birds	metamorphosis
bovine(s)	migrations
canine(s)	mollusks
carnivore	monotremes
cell(s)	omnivore(s)
cetaceans (Cetacea)	order
cold-blooded	organ(s)
crustaceans (Crustacea)	phylum (plural is phyla)
eggs	pinnipeds (Pinnipedia)
endangered	predator
equines (Equus)	Primates
exoskeleton	quadruped(s)
extinct	reptiles (Reptilia)
family (-ies)	rodents (Rodentia)
felines (Felidae)	species
fish	taxonomy
genus (plural is genera)	Vertebrata (vertebrates)
gills	warm-blooded

© Barbara M. Peller

Additional Terms

Choose as many terms as you would like and write them in the squares.
Repeat each as desired. Cut out the squares and randomly
distribute them to the class.
Instruct the students to place the square on the center space of their card.

Animal Characteristics Bingo

© Barbara M. Peller

Clues for Additional Terms

Write two or three clues for each new term.

_____

1.

2.

3.

_____

1.

2.

3.

_____

1.

2.

3.

_____

1.

2.

3.

_____

1.

2.

3.

_____

1.

2.

3.

© Barbara M. Peller

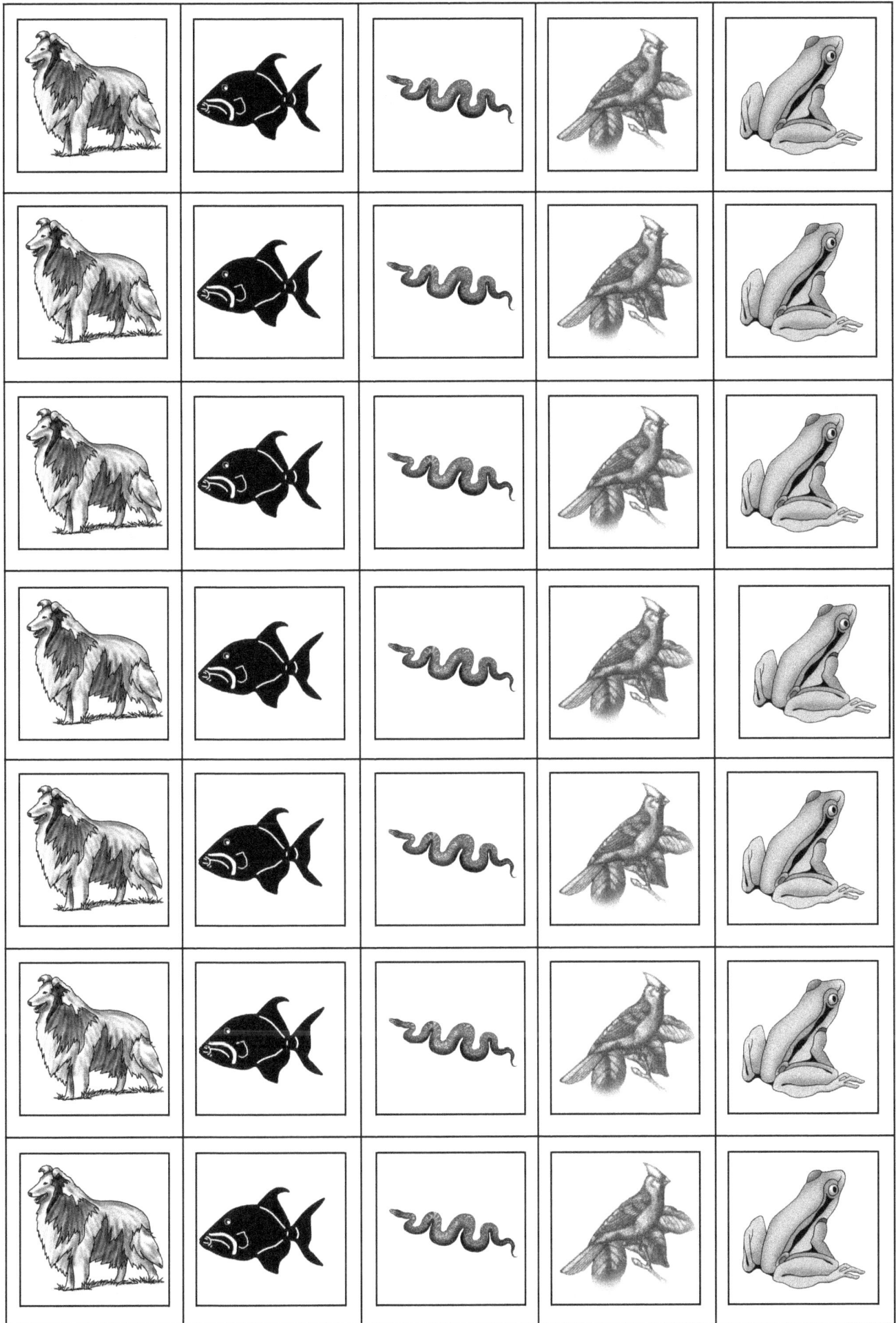

adaptation 1. An organism's changes to suit its environment is called ___. 2. An organism's ___ to suit the environment may be in its behavior, its physiology or its structure.	**amphibians** 1. Frogs and salamanders are ___. 2. Most ___ undergo complete metamorphosis. Their young have gills.
Animalia 1. Insects and spiders as well as fish, amphibians , reptiles, mammals and birds all belong to this kingdom. 2. Members of this kingdom are multicellular. They are heterotrophs, meaning they must get their food by eating other organisms.	**annelids (Annelida)** 1. Segmented worms are members of this phylum. 2. Earthworms and leeches are among the approximately 15,000 species of ___.
arachnids (Arachnida) 1. Spiders, scorpions, mites and ticks are ___. 2. Spiders and other___ have 4 pairs of segmented legs and a two-part body.	**arthropods (Arthropoda)** 1. The name of this phylum comes from two Greek words: *arthron,* meaning "joint," and *podos,* meaning "foot." 2. They have jointed limbs, a segmented body and an exoskeleton made of chitin. Insects, crustaceans and arachnids are ___.
biped(s) 1. An animal that walks on two feet is said to be a ___. 2. Humans are called ___ because they are two footed.	**birds** 1. These feathered animals are warm-blooded, egg-laying vertebrates of the class Aves. 2. They are the only class of animals to have feathers. Their forelimbs are modified to form wings.
bovine(s) 1. We use the adjective ___ to describe anything related to cows, oxen and buffalo. 2. Cattle and other ruminant animals are called ___.	**canine(s)** 1. Dogs, wolves, coyotes and foxes are members of the ___ family. 2. Dogs and other ___ are carnivorous.

© Barbara M. Peller

carnivore 1. An animal that eats meat is called a ___. 2. The term comes from two Latin words: *carne,* meaning "flesh"and *vorare,* meaning "to devour."	**cell(s)** 1. It is the basic unit of life. All organisms are made up of ___. 2. In the center of an animal ___is a nucleus. Its contents, called the protoplasm, is surrounded by a membrane.
cetaceans (Cetacea) 1. This order includes sea mammals such as whales, dolphins and porpoises. 2. The largest animal that ever lived, the blue whale, is a member of this order.	**cold-blooded** 1. Animals whose body temperatures are not regulated internally are said to be ___. 2. Reptiles and amphibians are said to be ___ because their body temperatures change according to their environment.
crustaceans (Crustacea) 1. This subphylum of arthropods has about 52,00 described species, including crabs, lobsters, shrimp. 2. Like most ___, barnacles live in the ocean.	**eggs** 1. Reptiles, birds and monotremes lay ___ with a hard, protective shell. They lay them outside of water. 2. Amphibians and fish lay their ___ in the water. Their ___ do not have a shell.
endangered 1. A species that is in danger of extinction is said to be ___. 2. Poaching for ivory has caused the African elephant to become an ___ species. We should protect them.	**equines (Equus)** 1. Members of this genus include horses, donkeys and zebras. 2. Horses and other ___ are medium to large mammals with long heads and necks and slender legs. Their necks have a mane.
exoskeleton 1. Humans and other vertebrates have an endoskeleton. Arthropods have an ___. 2. The hard outer structure that provides support and protection for insects and other arthropods is called an ___.	**extinct** 1. A species that is ___ no longer exists. 2. Dinosaurs are ___. So are prehistoric mammals such as the saber-toothed cat and the woolly mammoth.

Animal Characteristics Bingo

© Barbara M. Peller

family (-ies) 1. In the classification of animals, the ___ ranks above a genus and below an order. 2. Canidae (dogs) and Felidae (cats) are two ___ within the order Carnivora.	**felines (Felidae)** 1. This family includes lions and tigers as well as domestic cats. 2. Leopards and ocelots are ___.
fish 1. These limbless aquatic vertebrates are covered with scales and breathe through gills all their lives. 2. Bony ___ are the most diverse and numerous of all vertebrates. Sharks are cartilaginous ___.	**genus** 1. This taxonomic ranking is above species but below family. 2. Canis is the ___ that includes dogs and wolves.
gills 1. ___ are organs of respiration; their function is to extract oxygen from water. 2. The respiratory organs of fish and of the aquatic larva stage of amphibians are called ___.	**herbivores** 1. Animals that get all of their energy from eating plants are called ___. 2. Deer are ___. Dogs, which eat meat, are not.
insect 1. An ___ is an arthropod whose adult stage has three pairs of legs. 2. An adult ___'s body is segmented into three parts: a head, a thorax, and an abdomen. Most have two pairs of wings.	**invertebrate(s)** 1. An ___ is an animal without a backbone. 2. Insects and spiders are called ___ because they have no backbone.
kingdom 1. It is the highest ranking after domain in biological taxonomy. The next highest is phylum. 2. Animalia is the ___ comprising all living and extinct animals.	**larva (larvae)** 1. It is the wormlike stage of many insects before metamorphosis. The ___ of butterflies are called caterpillars. 2. It is the immature form of an animal before metamorphosis. This stage of a frog is called a tadpole.

© Barbara M. Peller

mammals (Mammalia)	marsupials
1. Animals belonging to this class are warm-blooded vertebrates with a covering of hair. 2. Females in this class have milk-producing glands with which to nourish their young.	1. Babies of these pouched animals are born in an undeveloped state. 2. Kangaroos, wallabies and koalas and other ___ differ from other mammals because the females carry their young in a pouch.
metamorphosis 1. Butterflies and moths undergo complete ___, which comprises 4 stages: egg, larva, pupa, and adult. 2. Dragonflies undergo incomplete ___, which comprises 3 stages: egg, larva and adult.	**migrations** 1. Many birds make seasonal journeys known as ___ in response to changes in weather, habitat and the availability of food. 2. The paths taken by birds during their seasonal ___ are known as flyways.
mollusks 1. Snails, clams, mussels, squid, and octopuses are examples of sea invertebrates known as ___. 2. ___ are invertebrates with a soft unsegmented body, usually enclosed in a shell; they are sometimes called shellfish.	**monotremes** 1. ___ are egg-laying mammals. 2. Platypuses and echidnas are ___.
omnivores 1. ___ get their food energy from both plants and animals. 2. Most humans are ___. They eat both plants and animals.	**order** 1. It is the taxonomic classification that ranks above the family and below the class. 2. Humans are members of the ___ Primates.
organ(s) 1. A collection of tissues which perform a certain function is called an ___. They may be organized into a system. 2. Skin, the heart and the brain are three ___ found in most animals.	**phylum** 1. ___ is the taxonomic rank that comes after kingdom and before class. 2. All vertebrates are members of the ___ Chordata. They belong to the sub___ Vertebrata.

Animal Characteristics Bingo

© Barbara M. Peller

pinnipeds (Pinnipedia)	**predator**
1. Members of this suborder of aquatic carnivorous mammal have a stream-lined body and flippers.	1. An animal that hunts and eats other animals is a ___.
2. Seals and walruses are ___. They use their flippers to swim.	2. The animal that is hunted and eaten by a ___ is called the prey.
Primates	**quadruped(s)**
1. This taxonomic order includes humans and monkeys.	1. An animal that walks on four legs is called a ___.
2. Members of the order ___ have advanced binocular vision, specialized appendages for grasping, and enlarged cerebral hemispheres.	2. Although humans have four limbs, only two are used for walking; therefore, humans are not considered ___.
reptiles (Reptilia)	**rodents (Rodentia)**
1. Members of this class breathe air and are cold blooded. Their skin is covered with scales.	1. This order includes animals with a pair of continuously growing incisors.
2. Snakes, lizards, crocodiles, and turtles are ___.	2. Mice, squirrels, beavers and hamsters are all members of this order.
species	**taxonomy**
1. A genus contains one or more ___.	1. ___ is the science of classification.
2. A group of organisms capable of interbreeding and producing fertile offspring are members of the same ___.	2. In biological ___ the major ranks of animals are kingdom, phylum, class, order, family, genus and species.
Vertebrata (vertebrates)	**warm-blooded**
1. ___ include fish, amphibians, reptiles, birds, and mammals.	1. ___ animals keep their body temperature at a relatively constant level without regard to the temperature of their surroundings.
2. This subphylum comprises chordates with backbones, or spinal columns.	2. Mammals and birds are ___. They maintain a stable internal temperature.

Animal Characteristics Bingo

© Barbara M. Peller

Animal Characteristics Bingo

exoskeleton	extinct	fish	species	quadruped(s)
bovine(s)	Animalia	rodents (Rodentia)	invertebrate(s)	family (-ies)
phylum	migrations		felines (Felidae)	larva (larvae)
taxonomy	amphibians	endangered	warm-blooded	genus
gills	Vertebrata (vertebrates)	carnivore	adaptation	equines (Equus)

© Barbara M. Peller

Animal Characteristics Bingo

taxonomy	predator	herbivores	metamorphosis	gills
genus	invertebrate(s)	birds	amphibians	organ(s)
monotremes	Vertebrata (vertebrates)		cell(s)	endangered
eggs	omnivores	migrations	mammals (Mammalia)	family (-ies)
equines (Equus)	rodents (Rodentia)	carnivore	bovine(s)	adaptation

© Barbara M. Peller

Animal Characteristics Bingo

taxonomy	endangered	invertebrate(s)	warm-blooded	phylum
Vertebrata (vertebrates)	Animalia	arthropods (Arthropoda)	extinct	kingdom
amphibians	rodents (Rodentia)		order	annelids (Annelida)
migrations	monotremes	gills	eggs	herbivores
adaptation	bovine(s)	carnivore	mammals (Mammalia)	fish

© Barbara M. Peller

Animal Characteristics Bingo

migrations	order	fish	bovine(s)	gills
insect	birds	extinct	metamorphosis	phylum
felines (Felidae)	eggs		quadruped(s)	warm-blooded
endangered	crustaceans (Crustacea)	rodents (Rodentia)	carnivore	arthropods (Arthropoda)
adaptation	equines (Equus)	marsupials	cold-blooded	larva (larvae)

© Barbara M. Peller

Animal Characteristics Bingo

equines (Equus)	quadruped(s)	amphibians	birds	bovine(s)
insect	endangered	arthropods (Arthropoda)	migrations	Animalia
predator	larva (larvae)		cetaceans (Cetacea)	fish
family (-ies)	order	exoskeleton	mammals (Mammalia)	cold-blooded
invertebrate(s)	carnivore	mollusks	cell(s)	felines (Felidae)

© Barbara M. Peller

Animal Characteristics Bingo

annelids (Annelida)	order	herbivores	predator	larva (larvae)
warm-blooded	amphibians	cold-blooded	extinct	phylum
metamorphosis	arthropods (Arthropoda)		birds	cell(s)
carnivore	gills	mammals (Mammalia)	marsupials	felines (Felidae)
genus	endangered	exoskeleton	mollusks	fish

© Barbara M. Peller

Animal Characteristics Bingo

exoskeleton	order	pinnipeds (Pinnipedia)	cetaceans (Cetacea)	invertebrate(s)
genus	fish	Vertebrata (vertebrates)	Animalia	insect
herbivores	warm-blooded		cell(s)	arachnids (Arachnida)
migrations	eggs	phylum	taxonomy	monotremes
carnivore	bovine(s)	mammals (Mammalia)	marsupials	annelids (Annelida)

Animal Characteristics Bingo: Card No. 7

© Barbara M. Peller

Animal Characteristics Bingo

felines (Felidae)	order	biped(s)	warm-blooded	arachnids (Arachnida)
insect	predator	metamorphosis	fish	quadruped(s)
phylum	organ(s)		larva (larvae)	birds
adaptation	migrations	taxonomy	cold-blooded	eggs
rodents (Rodentia)	carnivore	marsupials	amphibians	genus

© Barbara M. Peller

Animal Characteristics Bingo

cell(s)	invertebrate(s)	Vertebrata (vertebrates)	phylum	larva (larvae)
cold-blooded	predator	felines (Felidae)	amphibians	fish
kingdom	exoskeleton		Animalia	biped(s)
arachnids (Arachnida)	equines (Equus)	gills	cetaceans (Cetacea)	pinnipeds (Pinnipedia)
eggs	mammals (Mammalia)	arthropods (Arthropoda)	taxonomy	quadruped(s)

© Barbara M. Peller

Animal Characteristics Bingo

taxonomy	species	birds	metamorphosis	mollusks
larva (larvae)	arachnids (Arachnida)	extinct	Animalia	fish
order	organ(s)		warm-blooded	monotremes
gills	family (-ies)	cold-blooded	mammals (Mammalia)	kingdom
canine(s)	genus	herbivores	equines (Equus)	felines (Felidae)

Animal Characteristics Bingo: Card No. 10

© Barbara M. Peller

Animal Characteristics Bingo

annelids (Annelida)	organ(s)	amphibians	cold-blooded	genus
biped(s)	kingdom	cetaceans (Cetacea)	cell(s)	extinct
insect	predator		herbivores	Vertebrata (vertebrates)
canine(s)	phylum	mammals (Mammalia)	bovine(s)	taxonomy
arthropods (Arthropoda)	carnivore	exoskeleton	marsupials	invertebrate(s)

© Barbara M. Peller

Animal Characteristics Bingo

invertebrate(s)	eggs	kingdom	warm-blooded	cell(s)
Vertebrata (vertebrates)	rodents (Rodentia)	predator	marsupials	insect
exoskeleton	pinnipeds (Pinnipedia)		larva (larvae)	metamorphosis
carnivore	quadruped(s)	fish	taxonomy	Animalia
organ(s)	biped(s)	order	arthropods (Arthropoda)	arachnids (Arachnida)

Animal Characteristics Bingo: Card No. 12

© Barbara M. Peller

Animal Characteristics Bingo

canine(s)	quadruped(s)	annelids (Annelida)	kingdom	larva (larvae)
predator	biped(s)	order	cell(s)	monotremes
warm-blooded	birds		Vertebrata (vertebrates)	pinnipeds (Pinnipedia)
felines (Felidae)	mammals (Mammalia)	arachnids (Arachnida)	organ(s)	taxonomy
carnivore	family (-ies)	marsupials	exoskeleton	cetaceans (Cetacea)

Animal Characteristics Bingo: Card No.13

© Barbara M. Peller

Animal Characteristics Bingo

bovine(s)	predator	amphibians	cell(s)	canine(s)
arachnids (Arachnida)	exoskeleton	kingdom	Animalia	monotremes
cold-blooded	warm-blooded		herbivores	arthropods (Arthropoda)
family (-ies)	mammals (Mammalia)	order	birds	annelids (Annelida)
carnivore	metamorphosis	organ(s)	genus	felines (Felidae)

© Barbara M. Peller

Animal Characteristics Bingo

cetaceans (Cetacea)	cell(s)	amphibians	invertebrate(s)	fish
annelids (Annelida)	mollusks	extinct	predator	cold-blooded
larva (larvae)	exoskeleton		phylum	warm-blooded
carnivore	kingdom	biped(s)	mammals (Mammalia)	canine(s)
genus	eggs	marsupials	herbivores	Vertebrata (vertebrates)

© Barbara M. Peller

Animal Characteristics Bingo

birds	reptiles (Reptilia)	biped(s)	mollusks	omnivores
metamorphosis	organ(s)	pinnipeds (Pinnipedia)	insect	species
canine(s)	quadruped(s)		larva (larvae)	Vertebrata (vertebrates)
migrations	arachnids (Arachnida)	carnivore	cetaceans (Cetacea)	taxonomy
cold-blooded	kingdom	marsupials	eggs	monotremes

Animal Characteristics Bingo: Card No. 16

© Barbara M. Peller

Animal Characteristics Bingo

canine(s)	Primates	crustaceans (Crustacea)	kingdom	bovine(s)
cetaceans (Cetacea)	cold-blooded	mammals (Mammalia)	warm-blooded	pinnipeds (Pinnipedia)
cell(s)	taxonomy		reptiles (Reptilia)	biped(s)
equines (Equus)	genus	felines (Felidae)	amphibians	monotremes
gills	arthropods (Arthropoda)	invertebrate(s)	species	quadruped(s)

© Barbara M. Peller

Animal Characteristics Bingo

fish	order	arachnids (Arachnida)	cold-blooded	metamorphosis
equines (Equus)	canine(s)	amphibians	larva (larvae)	arthropods (Arthropoda)
cell(s)	monotremes		crustaceans (Crustacea)	mollusks
organ(s)	extinct	mammals (Mammalia)	taxonomy	herbivores
reptiles (Reptilia)	kingdom	gills	Primates	annelids (Annelida)

© Barbara M. Peller

Animal Characteristics Bingo

larva (larvae)	annelids (Annelida)	kingdom	biped(s)	organ(s)
cetaceans (Cetacea)	bovine(s)	birds	invertebrate(s)	species
Primates	warm-blooded		Animalia	mollusks
herbivores	reptiles (Reptilia)	gills	eggs	crustaceans (Crustacea)
phylum	omnivores	genus	felines (Felidae)	marsupials

© Barbara M. Peller

Animal Characteristics Bingo

organ(s)	Primates	species	kingdom	Animalia
birds	Vertebrata (vertebrates)	insect	gills	metamorphosis
quadruped(s)	pinnipeds (Pinnipedia)		migrations	crustaceans (Crustacea)
equines (Equus)	felines (Felidae)	adaptation	eggs	reptiles (Reptilia)
endangered	rodents (Rodentia)	omnivores	taxonomy	extinct

Animal Characteristics Bingo: Card No. 20

© Barbara M. Peller

Animal Characteristics Bingo

cetaceans (Cetacea)	annelids (Annelida)	insect	kingdom	family (-ies)
quadruped(s)	crustaceans (Crustacea)	arachnids (Arachnida)	biped(s)	exoskeleton
monotremes	genus		Primates	amphibians
gills	invertebrate(s)	reptiles (Reptilia)	equines (Equus)	felines (Felidae)
migrations	omnivores	marsupials	canine(s)	eggs

© Barbara M. Peller

Animal Characteristics Bingo

phylum	herbivores	crustaceans (Crustacea)	predator	canine(s)
metamorphosis	species	fish	biped(s)	Animalia
arachnids (Arachnida)	warm-blooded		exoskeleton	pinnipeds (Pinnipedia)
reptiles (Reptilia)	equines (Equus)	eggs	extinct	bovine(s)
omnivores	arthropods (Arthropoda)	Primates	monotremes	insect

© Barbara M. Peller

Animal Characteristics Bingo

birds	Primates	invertebrate(s)	predator	marsupials
annelids (Annelida)	organ(s)	genus	cetaceans (Cetacea)	extinct
herbivores	canine(s)		adaptation	exoskeleton
monotremes	rodents (Rodentia)	reptiles (Reptilia)	arthropods (Arthropoda)	eggs
family (-ies)	felines (Felidae)	omnivores	gills	crustaceans (Crustacea)

© Barbara M. Peller

Animal Characteristics Bingo

birds	organ(s)	bovine(s)	Primates	biped(s)
larva (larvae)	marsupials	insect	metamorphosis	exoskeleton
pinnipeds (Pinnipedia)	mollusks		canine(s)	monotremes
family (-ies)	adaptation	reptiles (Reptilia)	arthropods (Arthropoda)	quadruped(s)
endangered	migrations	omnivores	species	rodents (Rodentia)

© Barbara M. Peller

Animal Characteristics Bingo

migrations	insect	Primates	amphibians	crustaceans (Crustacea)
extinct	family (-ies)	cetaceans (Cetacea)	birds	Animalia
quadruped(s)	biped(s)		adaptation	reptiles (Reptilia)
mollusks	equines (Equus)	rodents (Rodentia)	omnivores	species
marsupials	bovine(s)	arachnids (Arachnida)	cold-blooded	endangered

© Barbara M. Peller

Animal Characteristics Bingo

crustaceans (Crustacea)	Primates	adaptation	metamorphosis	mollusks
herbivores	warm-blooded	biped(s)	organ(s)	birds
family (-ies)	gills		species	migrations
canine(s)	predator	equines (Equus)	omnivores	reptiles (Reptilia)
pinnipeds (Pinnipedia)	cold-blooded	amphibians	rodents (Rodentia)	endangered

© Barbara M. Peller

Animal Characteristics Bingo

adaptation	arachnids (Arachnida)	Primates	organ(s)	Vertebrata (vertebrates)
family (-ies)	herbivores	cetaceans (Cetacea)	reptiles (Reptilia)	Animalia
mammals (Mammalia)	rodents (Rodentia)		omnivores	migrations
mollusks	annelids (Annelida)	insect	endangered	extinct
canine(s)	species	crustaceans (Crustacea)	phylum	pinnipeds (Pinnipedia)

Animal Characteristics Bingo: Card No. 27

© Barbara M. Peller

Animal Characteristics Bingo

larva (larvae)	order	taxonomy	Primates	arachnids (Arachnida)
Vertebrata (vertebrates)	crustaceans (Crustacea)	adaptation	gills	species
rodents (Rodentia)	monotremes		mollusks	metamorphosis
pinnipeds (Pinnipedia)	phylum	genus	omnivores	reptiles (Reptilia)
predator	cell(s)	canine(s)	endangered	family (-ies)

© Barbara M. Peller

Animal Characteristics Bingo

crustaceans (Crustacea)	order	mollusks	cetaceans (Cetacea)	cell(s)
family (-ies)	gills	insect	pinnipeds (Pinnipedia)	phylum
quadruped(s)	Primates		Animalia	adaptation
Vertebrata (vertebrates)	equines (Equus)	fish	omnivores	reptiles (Reptilia)
birds	biped(s)	endangered	annelids (Annelida)	rodents (Rodentia)

© Barbara M. Peller

Animal Characteristics Bingo

bovine(s)	Primates	metamorphosis	cell(s)	reptiles (Reptilia)
extinct	mollusks	herbivores	species	Animalia
endangered	arthropods (Arthropoda)		pinnipeds (Pinnipedia)	insect
family (-ies)	annelids (Annelida)	order	omnivores	adaptation
equines (Equus)	invertebrate(s)	rodents (Rodentia)	crustaceans (Crustacea)	fish

Animal Characteristics Bingo: Card No. 30

© Barbara M. Peller

www.ingramcontent.com/pod-product-compliance
Lightning Source LLC
Chambersburg PA
CBHW051428200326
41520CB00023B/7392